Martha Lucía Posso Franco

The creation universal biology

Martha Lucía Posso Franco

The creation universal biology

Finally the biology of creation is complete!

ScienciaScripts

CREATION, BIOLOGY

UNIVERSAL INTEGRAL

MARTHA LUCIA POSSO FRANCO

CARTAGO AUGUST 31, 2016

Table of contents

INTRODUCTION : CREATION, UNIVERSAL INTEGRAL BIOLOGY

Long ago I taught you that the Creator and the creation are the same being. In this text I am going to show you what this marvelous integral being, God, is really like in his magnificent and marvelous creation.

We will see how each component or being of creation is endowed with a wonderful machinery, which makes us all fit into a wonderful whole, which functions as a single body and a single integral being. February 7, 2015.

MARTHA LUCIA POSSO FRANCO

AUTHOR

CHAPTER 1: CREATION, UNIVERSAL INTEGRAL BIOLOGY

Everything that is and exists materially, mentally and spiritually and the infinite universe: creation, is a magnificent being that changes its parts through its continuous transformation. It reinvents and keeps on self-creating itself, every second of the day, in each of its micronanoatoms and beings, infinite number of times, without fatigue and without delay, with the same accelerated rhythm since the beginning of life and creation.

CHARACTERISTICS OF THE CREATION

They are of two kinds: intrinsic and extrinsic. The characteristics of creation are the qualities it possesses, in relation to itself and its internal components and in relation to the nothingness and obscurity that surrounds it, or the space to be occupied with a material body.

Next, we will see how each of its characteristics is manifested.

INTRINSIC CHARACTERISTICS are all the internal qualities, feelings and integral behavior of the Whole and its parts, in internal feedback.

Among its intrinsic characteristics we have:

BEHAVIORAL, OR BEHAVIORAL. They are the cyclic or rotary movement, linear feedback and inverse. Those of feeling and internal sensory.

EXTRINSIC CHARACTERISTICS: Its relationship with the physical space that remains to be filled with material beings.

They are all the external qualities, feelings and integral behavior of the Whole and its parts, in relation to the nothingness or the space to be filled in the void that still exists in the infinite universe.

Among its extrinsic characteristics we have:

The physical ones in general that are part of each of the beings and stars of creation.

CYCLES OF THE UNIVERSE AND ITS BEINGS

ach one of the beings and stars of creation has its own life cycle from the moment they are born, transform into adults, bear fruit and age until they die and transform into dust that helps to generate and nurture new beings with their life cycles.

Just as each one of the beings and stars, the entire infinite universe also has the same life cycle: it was born, it is growing, transforming, maturing and bearing fruit in each of its components, from the tiniest to the macrogiant like supernovae and every day it ages little by little, until it shrinks completely, its physical body disappears and only a nanomicroparticle of energy remains, that in due time, when it matures, will explode in a new nanomicrobigbang and will begin the growth of a new material universe, which will fulfill in a lapse of megillions of years, the same life cycle of this one in which we live, until it finishes, matures again and propagates infinitely.

ENERGY IN GENERAL

On the planet and in the universe there is energy in all its forms: material, spiritual and mental. According to the nature of beings it can be: benign, malignant or neutral.

1. BENIGN ENERGY: when it tends to heal, evolve and improve the conditions, quality of life and habitat of all beings and stars of the universe.
2. EVIL ENERGY: when it tends to damage, sicken, involute and worsen the conditions, quality of life and the internal and external habitat of all beings and stars.

According to its origin it can be: material, spiritual and mental.

A. MATERIAL ENERGY. That which can be seen, touched and tasted; or with which one can have constant visual and physical contact, product of the psychophysical and esoteric feedback that accompanies it throughout its life process, from its birth to the death of the being and its transformation into dust and latent inert energy.

B. SPIRITUAL ENERGY. That which can be felt as a fluid, cannot be touched, and with great difficulty is perceptible to the human eye. It can only be seen by clairvoyants, witches, spiritualists, parapsychologists, spiritualists, Buddhists, gnostics and esoteric scholars in general. It can be found in its 3 forms:

BENIGN: when the feelings, the senses and the social fabric of the habitat of the area or being in question, is or are positive, benign and evolve constantly and progressively, in the improvement of each individual, their quality of life and total satisfaction and understanding, with the environment and all its beings. It tends to harmony and general peace.

MALIGN: when all the above are negative, malignant and involute constantly and progressively to the detriment of each individual, with the deterioration of their quality of life, misunderstanding, intolerance, detachment, hatred and chaos or total anarchy. A complete progressive disaster that affects the habitat and all its beings. It tends to war, mortality and destruction in general.

NEUTRAL: when inertia, apathy, conformity and freezing enter in the evolution of beings and habitat. The lazy and parasitic beings are part of this phenomenon, as well as the indifferent and disinterested ones. What is no longer cared for by inertia deteriorates and the habitat, the beings and the social fabric is non-existent or negative.

C. MENTAL ENERGY. That which cannot be seen, but is perceived in the balance or imbalance of beings or stars. Together with the material body and that of the spirit, it can produce phenomena such as telekinesis, mental dominion, hypnosis and mental or psychological regressions. It can be found in its 3 forms:

BENIGN: when it is totally controlled, it generates positive feelings and actions in beings and stars and their habitat in general. It can be appreciated but not touched. They are the positive thoughts and ideas and the positive acts or facts consequence of them.

MALIGNANT: when there is partial or total lack of control of the mind. It can be seen in psychic disorders: mild, regular or total severity, as in the insane. They can be bipolar disorders, schizophrenia, manias, phobias, vices, hatreds or total madness. In any of the cases, it is detrimental both for the one who is not well, and for those who surround him and his habitat.

NEUTRAL: when there are states of mental inertia such as: catatonia, mild coma or brain death. When they are suffered sporadically, as in lunatics or in those who collapse with regular epileptic seizures. They do not help to improve individuals, nor their habitat. They tend to deteriorate, not to improve the quality of life of beings and their habitat.

CHAPTER 2: SPECTRAL, SPIRITUAL AND MATERIAL PLANES OF THE UNIVERSE AND ITS BEINGS

SPECTRAL PLANE. The spectral plane of the universe is constituted by the negative or dark energies and just as the human being has a shadow, as well as the beings and stars, the whole universe has the sum of all the shadows of each one of its beings and stars, which conform it and when united, give origin to the spectral plane.

Therefore, those who, going against the natural laws of the universe and the laws of God, use it to fly and nourish themselves with their forms and knowledge, find in it a copy of all the beings and stars of creation. It is therefore this plane, the negative and deceptive aura of the universe. It has also been called up to now: the astral plane.

SPIRITUAL PLANE. It is intimately united to the material plane and all its manifestations of life, energy and force, are appreciable by the human eye. It is not found elsewhere, the only invisible parts of it are the energetic forms of beings who have left their material body and the celestial court, with the Trinity as supreme king.

MATERIAL PLANE. It is the physical plane, where everything that is and exists can be appreciated and touched with all the senses.

NOTE: the mind is present, administering each and every being, both in the material and spiritual planes, as well as in the plane of satan: the spectral or shadow of creation.

CHAPTER 3: SYSTEMS THAT MAKE UP EACH OF THE BEINGS OF CREATION

Like each and every being, the stars of creation and the entire universe have the same component systems. Among them are:

1. CIRCULATORY SYSTEM
2. DIGESTIVE SYSTEM
3. ELEMENTAL SYSTEM (VITAMINS AND NUTRIENTS)
4. ENDOCRINE SYSTEM
5. EXCRETORY SYSTEM
6. INTELLIGENT SYSTEM
7. MUSCULAR SYSTEM
8. NERVOUS SYSTEM
9. BONE SYSTEM
10. REAL OR MICTURATOR SYSTEM
11. REPRODUCTIVE SYSTEM
12. RESPIRATORY SYSTEM

Let us now see how each of them functions in the beings and stars and in the entire creation or universe.

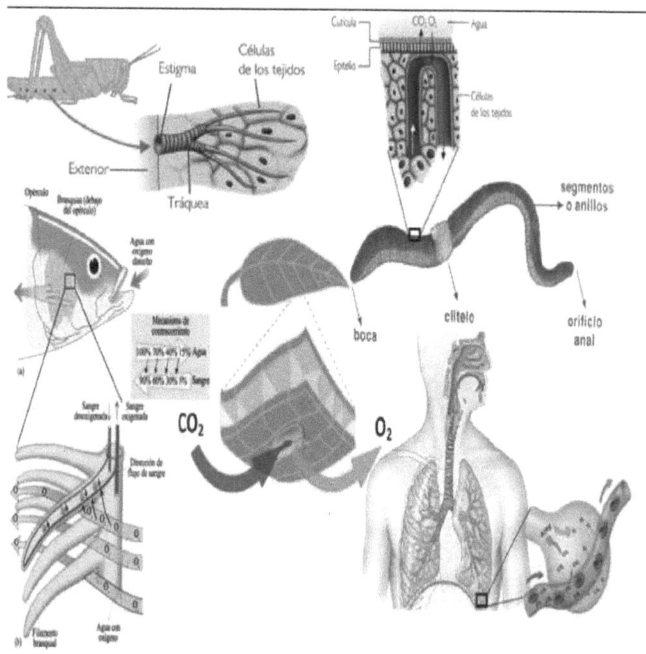

1. **CIRCULATORY SYSTEM.** Circulatory flow.

Circulating blood. It is composed of two movements or phases: diastole and systole: contraction movement of the heart and arteries. The main organ of this system is the heart, with whose beats the blood circulates at a normal rate through the body of the human being or animal.

a. IN ANIMALS. It is made up of veins, arteries and lymphatic vessels. By means of all its veins, it carries and extracts blood from the heart. It is the one that ensures that the human being can be alive and have a normal rhythm of life and movement.

b. IN PLANTS. It is the one that is formed by the conduits that transport the sap from the root to the tip of the tree or bush and returns it to the root, to follow the constant sap feedback cycle in the plant. It ensures life by circulating the sap, which is the life of the plant, through the vessels and cells that irrigate and keep it alive.

c. IN MINERALS. Liquids such as petroleum have this system, together with others such as the endocrine, excretory and respiratory systems; they help it to circulate the fuels, elements and ferments that form it to liquefy and produce the gases of the excretory system and to fulfill in a complex way, all the functions of the systems, in their nature. The metalloid or powdered minerals, have very slow these systems or lack them totally.

d. THE PLANET AND THE STARS. The planet's circulatory system is made up of all the planet's springs, tributaries, seas and oceans, freshwater, saltwater and wastewater.

The clean ones are those that enter to feed the planet and its beings. The dirty or residual ones, are the product of pollution or residues left in the waters by the beings of the planet. In the other stars and asteroids, which have any kind of liquid elements, the same thing happens.
All liquid elements are part of this system or the micturator, not only on the planet, but in all the stars and in creation.

The atmosphere with its cloud classes is an important part of this liquid circulation between it and the planet.

e. IN THE UNIVERSE. The circulatory system is made up of the water that circulates in the universe and the elements, in the form of gases or liquids. They move continuously and systematically. They are the blood of creation, circulating in the form of liquids, ice or gas. Driven by the movement, not perceptible to the human eye, of the heartbeat of the universe, coming from its enormous heart.

2. **DIGESTIVE SYSTEM.** The main organ of this system is the stomach. Food and liquids ingested through the action of saliva, gastric juice and bile are concentrated in it. They are digested and sent, according to their usefulness to the other systems and organs of the body, to continue their process of nutrition, processing or excretion among others. Related to digestion. System of the action of digesting.

   DIGESTION. Mechanical and chemical decomposition of food in the digestive tract, so that it can be assimilated by the organism: carbohydrates are converted into simple sugars, fats into fatty acids and glycol and proteins into amino acids, under the action of saliva and gastric juice.

   a. IN ANIMALS. It is formed by the esophagus and the stomach, where food and saliva enter. It works the gastric juice and bile. It is in charge of liquefying the food, separating the nutrients from the parts for the excretory or renal system and it is in charge of sending the feces to the colon and bladder and the toxins are expelled through the skin by means of sweat, after passing them through the liver to be filtered and sent to the kidney and bladder, to the skin or to the colon and anus. It ensures that the human being, can have energy to move and live with the nutrients and the expulsion of continuous feces.

b. IN PLANTS. As plants are static, their external movements are given by wind action. Internally they do have voluntary, continuous and systematic movements, which are not appreciated by the eye, the ear or the human or animal senses.

Just as it moves sap through the circulatory system, it also moves nutrients through liposomes and lysosomes (see botany). In the place that resembles the animal stomach, the food extracted from the soil or air is liquefied and the nutrients are sent to the whole plant, so that it can live and grow stronger, and the feces to the excretory system, where they are returned to the soil to be processed again and used for future plant nutrition. It ensures the life and nutrition of the plant and its duration as a living being.

c. IN THE MINERALS. It is a very slow system, by means of which, after a long period of time, with the substances that are arriving to the ground, it improves the mineral in some parts of the ground and in others it can impoverish it, until it becomes useless or disposable and in contact with liquids or water, they are washed and only the good mineral remains solid and the bad one is detached and separated from it.

It is important because it takes advantage of good minerals and powders and separates the bad and unusable ones. It can preserve or improve the quality of the mineral.

d. IN THE PLANET AND THE STARS. The digestive system of the planet is made up of all the solid, gas and liquid nutrients that feed back to each planet and star of creation. They are contributed by the same animal, vegetable, mineral and elemental beings that are part of each star and planet of creation. When the leaves or branches of vegetables fall, they decompose until they become dust, which enters to nourish, fertilize and increase the solid part of the planet, ensuring nutrients to generate millions of microorganisms and new beings, which regenerate the ecosystem and sustain the life, evolution and permanence of the being or star, for billions of years. The same happens with animal, mineral and elemental remains.

This digestive system is composed of the environment, the wind that spreads the dust and waste, the liquids that make it penetrate the earth of the star or planet and the time that, as it passes, with the help of the sun and the cold at night, contributes to the deterioration of solids and liquids and their enthronement in the soil, to give new life and new living beings on the physical plane.

e. IN THE UNIVERSE. It is made up of the large and small black holes that exist throughout the universe, which, with their centrifugal and centripetal forces, are responsible for taking what they reach, liquefying it and disappearing it into billions of micronanoatoms, which are scattered throughout the ether and go on to form new galaxies and solar systems.

CENTRIFUGAL FORCE. It moves beings or things away from the center of the black hole, tornado or whirlpool. It separates things or beings from the center of the atmospheric phenomenon.

CENTRIPETAL FORCE. It attracts towards the center of the atmospheric phenomenon, the beings and things it catches in its path.

In the black hole the centripetal force works by trapping stars, planets, beings or things; then these go through the worm body of the black hole, while they are digested or shattered into millions of nanoparticles, which upon reaching the other end, are taken by the centrifugal force and expelled far from the end of the black hole, towards the universe unknown now by the human being, to be used as components of new life in millions of new beings and stars, or discarded forever as dead and useless nanoparticles. The tornadoes or eddies act in the same way inside the solids and liquids of the stars or planets.

3. **ELEMENTARY SYSTEM.** Set of elements orderly related to each other.

ELEMENT. Each of the simple bodies which, alone or in

combination (compounds), constitute all known substances. The nature of each element determines the number of protons in its atomic nucleus; they are classified into metals and metalloids, about 105 elements are known, 92 natural and 13 produced by man.

a. IN ANIMALS. It consists of all metals, metalloids and liquids, other than water, blood, saliva, urine, feces and their irrigation system of the animal body.

b. IN PLANTS. Conformed by all the chemical elements and their irrigation system and composition of all the molecules of the plant body.

c. IN MINERALS. Solid minerals and liquids also have parts of chemical elements, with the corresponding irrigation system, which is slow in solids and more dynamic in liquids.

d. THE PLANET AND THE STARS. The elemental system of the universe with its beings and stars is made up of all the chemical elements that in greater or lesser quantity, are part of each being and star of creation, with its corresponding irrigation system, slower if it is solid and more dynamic if it is liquid.

e. IN THE UNIVERSE. Just as the elements are part of the composition of the planet earth, with its circulation or irrigation system and its other characteristics, they are found in the stars, asteroids and stars with the same characteristics, although in greater or lesser quantities and with better or poorer qualities.

Elements and minerals are found in the composition from nanomicroparticles to large supernovae. They are present in all beings and stars of creation.

4. **ENDOCRINE SYSTEM.** Set of glands that are orderly related to each other.

Glands that release their secretions (hormones) directly into the bloodstream, such as the thyroid, pituitary, adrenal, etc. (called glands of internal secretion), are called endocrine glands.

a. IN ANIMALS. It is conformed by the glands (organs that produce secretions that depending on which one they are, can be: endocrine, lacrimal, mammary, pineal, parotid, pituitary, prostatic, salivary, sweat, suprarenal or thyroid; in this case they are the endocrine as the name of the system indicates. Among them we have: thyroid, hypophysis, suprarenal, etc.

THYROID. Endocrine gland, located in front of the pharynx. It is also called the THYROID BODY: it secretes THYROXINE. Main cartilage of the larynx, its anterior face forms in man, the Adam's apple.

HYPOPHYSIS. Endocrine gland, located at the base of the skull called PITUITARIA.

SUPRARENAL. Endocrine glands covering the anteroposterior region of the kidneys, their medulla secretes adrenaline and their cortex elaborates several hormones, which discharge their secretions into the circulatory system. All these secretions are expelled residual substances which are then carried to the excretory system or exuded through the skin.

b. IN PLANTS. Consisting of all plant cells that help to exude toxins and residual substances from the plant.

c. IN MINERALS. They are liquid substances or powdered minerals, which are expelled or separated from the mineral to preserve their purity and original nature.

d. THE PLANET AND THE STARS. Conformed by the exudative, sweating and salivating system of all creation, which fulfills the cycle of evaporating the liquids, filtering them and returning them in the form of rain, according to its atmosphere and in the hard body of the star or planet, its liquids, are the residual, dirty or lethal substances that can be separated from the innocuous or benign ones, regardless of their nature: solid, gaseous or liquid.

5. **EXCRETORY SYSTEM.** Set of glandular organs, orderly related to each other, which excrete the glands through the excretory duct. System of organs responsible for carrying feces through an excretory duct out of the material body. The colon is a major part of this system.

COLON. Second portion of the large intestine, between the cecum and the rectum; includes the ascending, descending, transverse and sigmoid (or that iliac) colon.

BLIND. Part of the large intestine between the ileum and the colon.

RECTUM. Last portion of the large intestine.

ANUS. The final orifice of the rectum.

SIGMOID, SIGMOID. Similar in shape to sigma, s-shaped.

TRANSVERSE. Directed or placed across. Traversing from one side to the other.

a. IN ANIMALS. System consisting of the large intestine, rectum and anus; which varies according to the animal, from simple to complex organs. They expel all solid substances that are waste for the physical body.

b. IN PLANTS. Set of organs and excretory cells, orderly related to each other, responsible for expelling from plants, the solids that are not needed.

c. IN MINERALS. It is the mechanism of solids filtration, in the natural mineral purification process. This process can take from decades to millions of years.

d. THE PLANET AND THE ASTROS. As in the beings, every astro or asteroid, in each of its parts, filters liquids, solids and gases, expels at the end the residues of all these, which it does not need for its subsistence in solid form, in particles or in great masses. Part of this process are the solid substances expelled inside the lava of volcanoes. Also the vents that expel sulfur and other solid substances.

e. IN THE UNIVERSE. As in the beings and stars, the whole

universe has a well formed excretory system that unites each and every one of the excretions of the components of the universe. The atmospheric phenomena, are part of this system, when without any detectable physical origin, we feel odors expelled by the earth or the other planets, gases and all the residues and process of degradation of all physical matter in dust, are part of this excretory system of creation.

6. **INTELLIGENT SYSTEM.** (Discovered since 1999 by the author and finished in 2016, date in which I have published it in my blog: profetisamarthalucia.blogspot.com.co.

INTELLIGENCE: Faculty of understanding or knowing.

NERVE: Each of the whitish cords that radiate from the brain, spinal cord and other organs and transmit sensations and motor impulses: acoustic nerve, optic nerve, olfactory nerve, among others.

NEUROGLIA: Set of cells provided with long branched prolongations, which are located between nerve cells and nerve fibers in vertebrates.

PNEUMOGASTRIC: nerve extending from the medulla oblongata to the cavities of the thorax and abdomen.

NEURITE: A filiform extension that starts from the nerve cell and contacts muscle cells, glandular cells, etc., or the body of another nerve cell.

NEURONE: nerve cell.

CEREBRUM: upper and anterior part of the brain.

CEREBELLUM: posterior and lower part of the brain.

ENCEPHALon: set of nervous organs contained in the skull, comprising the cerebrum, cerebellum and medulla oblongata.

MERDULE: extension of the brain, which occupies the vertebral canal.

OBLONG MARINE: Anterior portion of the human spinal cord, which is considered part of the encephalon.

MAIN PARTS OF THE INTELLIGENT SYSTEM

The brain complete with neurons and their sensitive neurotransmitter filaments are the main components and the accessories are the nerve cells with their neurotransmitter fibers that are part of each chromosome, or organ of the human body.

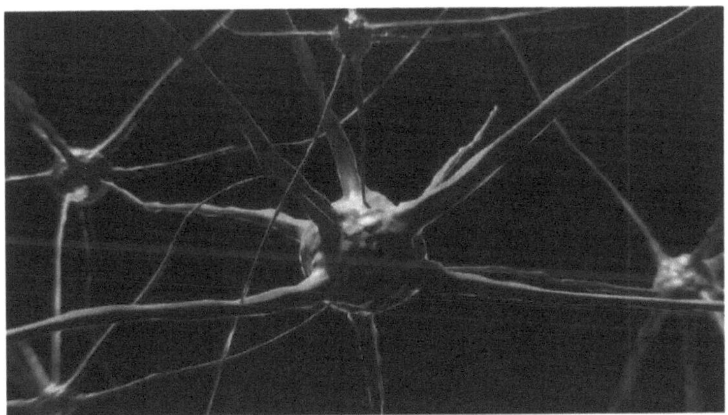

Neurons, the most important parts of this system.

INTELLIGENT SYSTEM, WHAT IS IT? It is the set of neurons with their whitish cords and transmitter networks, which are in all the cells and chromosomes of the human body. They are the ones that allow the being to see, perceive and understand who or what he is, with all his characteristics and to perceive, understand and be able to discern about himself and what surrounds him and identify him fully, to be able to react according to that knowledge and integral perception of himself and what surrounds him. It also allows him to access day by day to new perceptions and knowledge without stopping, it is cumulative. It works from the central brain to the auxiliary brains of each component of the human body and from those auxiliary brains located in the organs of the human body to the higher central brain.

a. IN ANIMALS. Consisting of all neurons that are part of the entire animal body with its fibers and extensions. They carry and bring the understanding of its internal and external environment from itself to everything around it.

b. IN PLANTS. Conformed by all the nerve cells with their prolongations and fibrous ramifications that are found throughout the plant body, carrying and bringing the knowledge or understanding of its inner essence and what surrounds it.

c. IN MINERALS. It is not easy to see to the human eye. It is what makes it react seeking its conservation, in contact with other beings or elements of the outside world, it is found in each of its atoms and molecules, similar in its functioning, to neurons.

It takes hundreds of thousands of years to notice this quality if it is not carefully observed.

e. THE PLANET AND THE STARS. Conformed by the sensitive fibers of atoms and molecules, which function just like neurons, it gives them the way to react, to conserve, evolve and strengthen themselves.

7. **MUSCULAR SYSTEM**. The set of muscles that cover the skeletal system of any animal being in the universe. They are the ones that produce the movements in the animal body.

a. IN ANIMALS. Made up of all the muscles that cover their skeletal system.

b. IN PLANTS. It is the tissue that makes up the body and branches of the plant, its flesh, or mass.

c. IN MINERALS. It is the main component that gives consistency and its own qualities to each mineral.

d. THE PLANET AND THE STARS. It is the part that covers and gives solidity and physical presence to every being, star, planet, planetoid or star in the cosmos.

8. **NERVOUS SYSTEM.** Set of nerves that are part of the being, and that are interconnected with each other, deliver and receive actions and reactions, sensations, senses and feelings from the main brain, its neurons and neurons or nerve cells of each of the organs of the body of the being, or atoms or sensitive chromosomes, of the body of the astro.

An important part of these neurons are their fibers and nerves. Each one of them are the transmitters and receivers that radiate the whole being, transmit sensations, actions and reactions. They are the ones that control all the senses.

a. IN ANIMALS. Set of nerve cells with their transmitter fibers, which carry and bring actions and reactions or sensations from the brain to the whole body and vice versa, from each of the organs and their nerve or neuronal endings, to the main brain.

In the main brain these nerve endings are in the neuronal fibers or neurons that are neurotransmitter-receptor and in the rest of the body in all the atoms and chromosomes of its neurons or nerve endings.

b. IN PLANTS. Fibrous bundle of the underside of the leaves, set of filaments that carry sensations throughout the plant body, from its root to the tip of the branches and vice versa, from each of the fibers of the organs of the plant, its nerve endings or neurons of the auxiliary brains of the plant. It is in the root, where the main brain of the plant is, contrary to the place where the main brain of every animal being is located.

c. IN MINERALS. It is found in the molecules of the mineral, whether solid or liquid, is what allows it to react to be, combined with other minerals or elements, by natural means or by the hand of man. To notice it, it must be carefully observed. This is the only way to appreciate all its metabolism and changes that evidence it, with the passage of time or in reaction to combination with other elements or external intrusions in its natural habitat.

d- IN THE PLANET AND THE STARS. As in the minerals, it may take a long time to notice or see the change or changes. The difference is that the whole structure in the stars is shaped by its own composition, in feedback with the beings that inhabit inside and outside of it.

9. **BONE SYSTEM.** Set of bones that make up the skeleton of the animal being, which are distributed throughout the body and serve to protect the internal organs and support the muscles and external part of the being. Its main components are the bones that make up the skeleton of the animal.

BONE: each of the hard parts of the skeleton or framework of vertebrates. They are formed of materials hardened with mineral salts (carbonates and phosphates of calcium and magnesium). They are endowed with nerves and blood and lymphatic vessels, between millions of microscopic longitudinal channels, which make them porous in the reddish part of the marrow, which is inside, the red blood cells are formed. Hard inner part of some fruits. Branches that hold together powders, rocks and solid mineral formations.

VERTEBRAL SPINE: backbone. Bone formation with intercalated vertebrae running from the head to the coccyx, supporting the skeletal framework of the animal's body, covered by muscles, nerves and skin.

SKELETON: bony framework of the animal body.

SPINAX: set of vertebrae running from the nape of the neck to the tailbone or coccyx.

vertebra: each of the bones that make up the spine, from the nape of the neck to the waist, from the waist to the rump, and in the upper and lower extremities.

a. IN ANIMALS. Set of bones that make up the skeleton of an animal. They are located in: the head or skull of the animal, the vertebral column is the one that supports it from the lower part of the skull or nape of the neck, to the coccyx or birth of the tail.

b. IN VEGETABLES. It is found in the structure that gives shape to the tree, formed by the main trunk with its internal hard component, the branches and the hard part or shell of the seeds and the interior of the fruits.

c. IN MINERALS. Branched stony configuration, which unifies and holds together solid minerals and rocks.

d. ON THE PLANET AND THE STARS. Configuration of calcic consistency, earthy and other solid elements, which branched, supports the solidity of the planet, astro rock or asteroid; united.

10. **RENAL SYSTEM.** Integrated set of organs that help to evacuate the urine and all liquid that is left over in the animal or vegetable being. It is called renal because it is related to the kidneys, which are its main organs, and are connected to the urethra and bladder.

KIDNEY: viscera that secretes urine.

URETRA: duct through which urine is expelled.

Ureter; duct leading from the kidneys to the bladder, through which urine passes.

VEJIGA: membranous sac containing urine.

a. IN ANIMALS. An integrated set of kidneys, ureter and bladder, which help to evacuate urine.

b. IN VEGETABLES. Set of pores or cells that help the vegetable to evacuate the excess liquids it no longer needs.

c. IN MINERALS. In porous minerals, it is the system of expulsion of excess substances.

d. IN THE PLANET AND THE STARS. Set of ducts and channels that help the planet or star to expel or dispose of excess or no longer needed waste liquids, such as those expelled in geysers and the liquids that are mixed in the lava of volcanoes. It is linked to the excretory system of the planet or star.

Set of liquids that are left over to each and every one of the animal, vegetable and mineral beings that make up the planet or star.

11. **REPRODUCTIVE SYSTEM.** The male and female reproductive organs in animals and plants. Its main organs are: the uterus or womb and the ovaries, in the female sex, and the spermatozoa, in the male sex, among humans and animals. In plants, they are: pollen, gametes, zygote, seed, ovaries and eggs or oocytes.

a. IN ANIMALS. Reproductive process, caused by the reproductive or reproductive organs.

MATRIX: uterus. Viscera of the female, in which the animal fetus grows.

OVARY: female reproductive organ that produces and contains eggs.

OVUM: sex cell contained in the ovaries of females.

Spermatozoa: male sex cell that fertilizes the egg.

Spermatozoa: spermatozoa of animals. Spermatozoa.

b. IN PLANTS. Generated in plant reproductive organs.

GAMETES: each of the male or female sex cells.

POLLEN: fertilizing powder contained in the ovaries. Part of the pistil that contains the ovules.

Ovule: sex cell contained in the ovaries of flowers.

CIGOT: fertilized ovum.
CYTOE: cell formed by the union of a male and a female gamete. Organism that develops from this cell.

SEED: part of the fruit of the plant, which contains the germ to reproduce it.

c. IN MINERALS. Generated by millenary evolution. Inorganic substances, (without organs for life). Useful part of the material extracted from a mine. They exist in powder, metals, metalloids and liquids.

Its reproduction is the product of a transformation process, generated by the evolutionary feedback of the planet's dust, with internal and external agents and elements. This process can last from thousands to millions of years.

d. IN THE PLANET AND THE UNIVERSE. In the stars, asteroids, planets and planetoids it is the same as in minerals. In the universe its main initial reproducer is the big bang, which resembles the bursting of the ovary of the flower when spreading the pollen.

In the same way, the outburst of a supernova or the primordial outburst or big bang is followed by the process of pollination, or in the case of stars, I will call it: the process of big banguization.

BIG BAN: process of universal fertilization and reproduction, produced by the primordial outburst and prolonged by subsequent or following big bangs, in perpetuity of a supernova or superstar or sun-like stars.

12. **RESPIRATORY SYSTEM.** Set integrated by the organs of respiration, whose main organs in animals are the lungs and in other species are the cells. Its main function is to breathe.

BREATHING: action and effect of breathing.

BREATHE: to absorb air from living things and expel it in a modified form. To absorb oxygen from water, fish. To exhale an odor. Physiological function that supplies oxygen to the cells of an organism and eliminates carbon dioxide. Breathing air.

LUNGS: respiratory organs of humans and air-breathing animals (vertebrates). Respiratory organ of certain arachnids and terrestrial mollusks.

PULMONATES: articulated animals that have lungs.

Nose: protruding part between the forehead and the mouth, with two orifices through which the animal takes in and expels air. Sense of smell. Organ through which the functions of breathing in and out are performed, it is interconnected with the bronchi or lungs.

BRONCHUS: each of the two ducts into which the trachea bifurcates and enters the lungs, where it divides and subdivides until it ends in the air vesicles.

BRONCHIOLUS: each of the small ducts into which the bronchi are subdivided in the lungs of animals.

a. IN ANIMALS. Respiratory process, generated by the uniform movement of the lungs, in their function of breathing in and out, in humans and animals.

b. IN PLANTS. Respiratory system, generated by the interconnected system of vesicles and respiratory cells. During the day they expel oxygen and inhale carbon dioxide, at night they inhale oxygen and expel carbon dioxide.

ASPIRING: to draw air into the lungs.

EXHALE: expel the aspirated air.

BREATHE: to absorb air from living things and expel it in a modified form. Absorb oxygen from water, fish.

c. IN MINERALS. The respiratory system of minerals is included in each atom of the mineral dust that composes it.

d. IN THE PLANET AND THE UNIVERSE. In the stars and planets it is the same as in the minerals, it is present with the circulation of the air outside and inside the body of the star, planet or planetoid and in each of the animal, vegetable and mineral beings that exist in it.

CHAPTER 4: EVOLUTION AND INVOLUTION

EVOLUTION

Action of evolving. Theory according to which the existing animal, vegetable, mineral species and stars (beings and stars) are descended from simpler organisms, which have been gradually modified from generation to generation.

a. IN ANIMALS. It is present since the beginning of animal life on earth, through the first living beings, the unicellular beings.

UNICELLULAR: beings made up of a single cell. After hundreds of years of evolution, multicellular beings appeared.

PLURICELLULAR: beings or organisms formed by many cells.

b. IN PLANTS. It is present from the appearance of single-celled organisms to multicellular organisms and continues to evolve.

c. IN MINERALS. It has been present since before the planet was formed, with the appearance of God's nano microparticle, the first burst of a primordial nanobig bang and the first cosmic dust until now and will continue into the future.

d. ON THE PLANET AND THE UNIVERSE. It is presented with the appearance of the first explosion of the first nanomicroparticle of God, with the first nanomicrobig bang. When the first one exploded, it scattered its atoms around and started its infinite reproduction system.

Atoms, chromosomes, cells were formed, until beings and stars were created and formed, ranging from unicellular to multicellular and supernovae.

Evolution is actually the changing process of fertilization of the entire cosmos or universe and each of its component beings and stars, whose primordial fertilizing function is given in the first nanomicrobig bang, until the megamacrobig bangs, which have occurred and will continue to occur in the future.

It is the renewal mechanism of the universe. Its main engine of fertilization is the Big bang.

INVOLUTION

Turn inward. Backward. Turn or reverse in the generative process of life.

a. IN ANIMALS. Some species have disappeared because instead of improving their metabolism and evolving their species in the face of generational changes in their habitat, they remain unevolved, do not adjust to the changes and become exhausted, exhausted, delayed or decayed, until they disappear.

b. IN VEGETABLES. As with animals, many species and families have disappeared because they have deteriorated to the point of disappearing and not even the seeds remain.

c. IN MINERALS. It is noticed with the deterioration and degeneration of the quality and hardness of solid minerals, or firmness in the consistency of liquids, until they become malleable, of little or no quality, and unusable.

d. IN THE PLANET AND THE UNIVERSE. Cycle by which the stars and planets reach old age, exhaustion and exhaustion of its context and tend to self-destruct or disappear by explosion or implosion.

EXPLOSION: action of bursting a body in the form of an explosion, with a roar. Sudden, violent combustion.

IMPLOSION: action of a star or planet breaking inward with a roar. Tendency to diminish the planet or astro little by little, consuming itself inward until it disappears and becomes cosmic dust. Abrupt decrease in the size of the astro, planet or planetoid, which is being sucked like a raisin that dries up and turns into dust. When the star or planet reaches this state, it is pulverized.

IMPLOSION

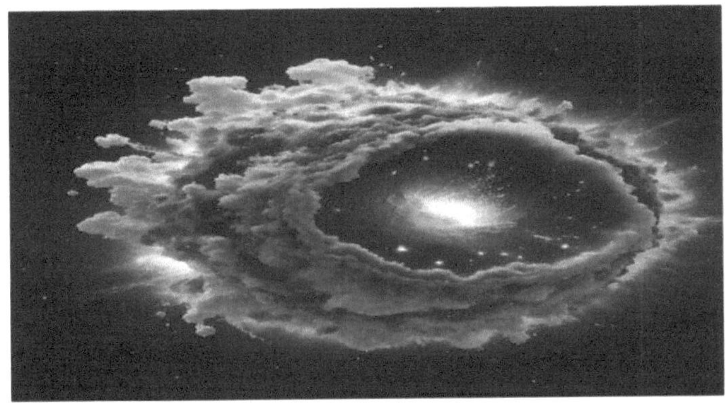

EXPLOSION

CHAPTER 5: DEATH

DEATH: cessation of life. Termination of the vital functions of any being or star of creation.

a. IN ANIMALS. All rational, semi-rational and irrational animal beings; from the moment they are born, every hour and day that passes, they begin to die. The death of each of these animal beings occurs for various reasons and circumstances, depending on the lifestyle they lead and the environment or habitat in which they reside.

When its material body dies, its soul and energy remain integrated into the Whole or creation. His material body becomes fertilizer and generation of life for millions of microorganisms that are born from his remains and finally becomes dust, which serves as nutrient for millions of specimens, until the mortal remains are completely exhausted, if the mortal remains are not stored or preserved in a special place.

b. IN VEGETABLES. They also age and die like animals. They fulfill their life cycle and when they die, their offal generates new life for others and for their offspring.

c. IN MINERALS. With the passage of the human being through the deposits of solid or liquid minerals, they disappear or are transformed. On their own, they last from thousands to millions of years, until the star or planet in which they are found evolves or devolves, until they disappear or are transformed.

d. IN THE STARS AND THE UNIVERSE. It is the exhaustion of life in the stars and planets. It can occur by evolution, transformation or involution. It can be given by slow process, with the course of the centuries or fast by intervention of agents or elements internal or external to it, that take it to explode or implode. They age, die and transform, just like the other beings and stars of creation.

CHAPTER 6 : TRANSFORMATION AND RECYCLING OF THE UNIVERSE AND ITS BEINGS

RECYCLE: to submit waste or garbage of any material or waste, purify or clean it and submit it with its peers to a process of recovery and change, to use it in other ways, for the benefit of all.

TRANSFORM: to change the form and usefulness of things or beings.

a. IN ANIMALS. During their material life, from their birth to their death, they undergo the changes proper to the stages, from growth to old age and death.

TRANSFORMATION: Once disembodied, it becomes energy and spiritual presence, pure energy, which can be good, static or neutral or bad, depending on the performance of the being in its material life and the habitat and influences that affected the being or astro.

The spirit passes to another plane, where it begins its evolution, until it becomes an advisor, guardian or guide of the material beings that it must help or take care of.

RECYCLING: Their material remains are recycled by nature and it is there where millions of microorganisms are born to renew the ecosystem, purify those remains and take full advantage of them as nutrients for the new beings that are born or reproduce from them.

b. IN VEGETABLES. They undergo the same cycle and process as animals, from birth to death.

TRANSFORMATION: already stripped of their vegetable body, they become energy with a spiritual body, which can be good, static or neutral or bad, according to their material nature, the use given to them and the habitat in which they existed.

His energy or spiritual body remains in the spiritual plane, providing the service of energetic accompaniment to other plants of the same nature or helping to harmonize with his spiritual presence, the place he occupied in his material life.

RECYCLING: Their material remains are recycled by nature until they are pulverized and converted into fertilizer for new plants. They also generate millions of new microorganisms that renew the habitat, with their new life cycles and reproduction.

c. IN MINERALS. They suffer the process of wear and tear that causes their disappearance and death, they are not renewable.

TRANSFORMATION: they need human beings, animals and vegetables, to transform themselves or their coexistence with other minerals, for their transformation into substances or mixed minerals.

The natural transformation occurs with the passage of time, until it disappears in the form of cosmic dust. Their energy or spiritual presence goes on to help harmonize the habitat where they existed and gives their presence on the spiritual plane, as harmony in biodiversity. They can be good, static or neutral, or bad.

RECYCLING: mineral wastes can be recycled or transformed with external help, or by manufacturing objects and equipment with them.

d. IN THE STARS AND THE UNIVERSE.

TRANSFORMATION: it is present from the creation of the first particle of God or God, until a megamacrosupernova. From the moment they are born until they die, they are constantly transformed, until their material form disappears by death, generating other beings or cosmic dust and transforming into new stars or planets.

Its energy or spirit, depending on whether it has been good, static, neutral or bad, goes on to help harmony in the space it occupied or to integrate itself into the spiritual plane, as part of it, with its full presence, for the good of the Whole and creation.

RECYCLING: nature is in charge of the change, evolution and transformation of cosmic dust, debris and astro debris, to generate new life and give birth to new stars, planets and planetoids that renew the universe.

It is a continuous process without pause, through which all beings and stars of creation must pass.

CHAPTER 7: RESURRECTION OF THE UNIVERSE AND ITS BEINGS TO THE SPIRITUAL OR ENERGETIC PLANE

RESUCITAR: To come back to life.

RESURRECTION: act of coming back to life who has been deceased. Action of resuscitating the beings and stars of creation because they are an integral part of creation, they are trinitarian: BODY, MIND AND SPIRIT. By that quality all die in the physical or material world or plane and pass in energetic body, conformed by mind and spirit, to the spiritual plane or world.

Once there, animals, plants and minerals, as well as the stars, serve as energetic support to the new material beings, which are born in the physical part of creation and are born in the physical part of creation.

The good natured ones act as advisors or guardians, or energetic nourishers of them. Those of a bad nature, when they are not cast into nothingness, are left as distracters, hinderers or obstructors of material beings.

a. IN ANIMALS. When their physical body dies, their soul: mind and spirit, goes to the spiritual plane to a period of rest and recovery. Once recovered, they pass to the groups of their genus, species and families to meet their relatives of their small family each one of them, then they pass to the training sector, where they are assigned their tasks as advisers, guardians or good energetic feeders.

The bad ones are left to wander some, trying to do more damage spiritually. Others go to a purgatory or zone where they must work and evolve, where they must try to repair the damage done on the physical plane during their material existence. Others go to the place of the wastes of creation, where the universal garbage is thrown, where they stay and never return.

b. IN VEGETABLES. When the plant dies, its material body becomes compost and its soul passes to the spiritual plane, to a period of recovery and adaptation, for its future work, as support of those of the same species and family, in the spiritual plane, those of good nature. Those of bad nature, pass to a place of transformation or spiritual recycling. Those who are transformed into good, go on to help, just like the good ones, those who do not want to change are discarded in the avernus or garbage dump of creation, where their essence is discarded forever.

c. IN MINERALS. When the mineral is exhausted or its physical presence is totally finished or transformed, its spiritual energy, after a period of rest and recycling, if it is good, goes to improve the habitat where it was and was exploited. If it is bad, it goes through a process of transformation and recycling of its spiritual energy, to then send it as an energetic enrichment of its habitat. If its essence is still evil, its spiritual essence is discarded in the dustbin of creation.

d. IN THE STARS AND THE UNIVERSE. When the physical body of a star disappears by explosion, implosion or exhaustion, its remains go on to give life to new physical beings by means of cosmic dust and its residues. Its remains are like pollen or seeds, which are carried or helped by the wind to pollinate or fertilize new areas of creation with their physical sprouts.

His soul, if it was good, goes to recovery and restoration of energies. Then it is trained and sent as an energetic aura that advises or helps other stars, asteroids, planets or planetoids, of the same species.

If its performance or behavior was bad, it is sent to a place of transformation or recycling, just as it happens with vegetables, animals and minerals, and if it cannot be transformed, its bad energy or bad spirit is discarded in the garbage dump of creation, forever.

CHAPTER 8: REBIRTH OR REBIRTH OF THE PHYSICAL UNIVERSE WITH ITS BEINGS

RELEVO: action of relieving one.

RELEASE: to exonerate from a burden, employment or commitment; absolve. To replace something or someone.

RENACER: to be born again.

REBROTAR: to sprout.

Sprout: plant shoot.

BROTAR: to sprout the plant from the ground. To give birth to shoots, leaves, etc.

The universe or creation is replaced second by second, from the emergence of the first nanomicrochispita or nanomicroparticle of God, which is the smallest of creation, to a supermegamacronova and simultaneously continues to grow and increase, its number of beings and stars, in a cycle of life, through the infinite and unstoppable relay of each of its atoms, molecules and parts that make up the universe or creation.

a. IN ANIMALS. Their material bodies when they die are offal from which millions of new beings arise, which in turn will give nutrients and food to the higher beings in the material food chain. Physical animals reproduce by the relationship between the two sexes, either by egg, by gestation or by partogenesis. All species reproduce and multiply, generation after generation.

Spiritually, one passes from physical death to spiritual life; it is the rebirth of the soul, when it returns to its place of origin.

b. IN PLANTS. Its remains serve as fertilizer or seedbed to generate millions of new beings, of other species and of the same species and genus, thus ensuring its relay, rebirth in its children or being replaced by them.

They are reborn on the spiritual plane, when they leave their physical body and return to their place of spiritual origin, to continue their life mission, as advisors, guardians or energizers.

c. IN MINERALS. Their residues go on to enrich the soil or impoverish it, with the help of animals. They disappear when they are extracted from their rock or soil and transformed into new products or discarded as useless or harmful in their physical body.

Its spiritual essence is reborn when the star or planet that contains it disappears or when it is completely extracted from its deposit or vein. It is reborn with the new stars that are formed in the generation of life released from creation.

d. IN THE STARS AND THE UNIVERSE. When they perish by explosion, implosion or depletion, the microparticles and physical debris that it waters give new life to galaxies, systems and new ecosystems in general.

In the spiritual realm, its energetic aura becomes its work of help or harm, depending on its good or bad nature, for the new physical beings that succeed it in the whole of creation.

AUTHOR'S NOTES

I have called it **THE CREATION, UNIVERSAL INTEGRAL BIOLOGY**, because I hope I have summarized all the spiritual, bodily and mental characteristics of the infinite universe and its beings and stars in this text. In addition, I have added the **INTELLIGENT SYSTEM**, which no scientist had yet discovered before me.

Knowledge enables us to take care of ourselves and others properly and helps us to take advantage of all the resources and bounties of creation, for the good of all in the Whole.

Once read, each of you, you already know the true essence and existence of the universal creation, with its beings and I started it on February 7, 2015 and finished it on November 1, 2016, at 8:30 PM.

MARTHA LUCÍA POSSO FRANCO

Author

BIBLIOGRAPHIC REFERENCES

The free photos and images, I got them from Google from Pixabay and some others, from the site's compendium.

The general biology was already done, the trinitarian biology of body, mind and spirit I added, thanks to my studies and research of more than two decades.

I hope you find this new knowledge useful.